Lah'aj Kids Division Worksheets
With Ṡabaṫiyya

By Tiyi Hibner
©2025 Yamartat Ta'/Tiyi Hibner
Published by www.TempleBabies.com

All Rights Reserved, including the right
to reproduce this book or portions thereof
in any form without permission from the publisher,
except as permitted by U.S. copyright law.

For permissions contact the publisher at:
info@TempleBabies.com

© 2025 TempleBabies LLC
All Rights Reserved.

Lah'aj Kids
DIVISION WORKSHEETS WITH ṠABATIYYA

ṠABAT·IYYA HARAF·AAT — SABAEIC ALPHABET

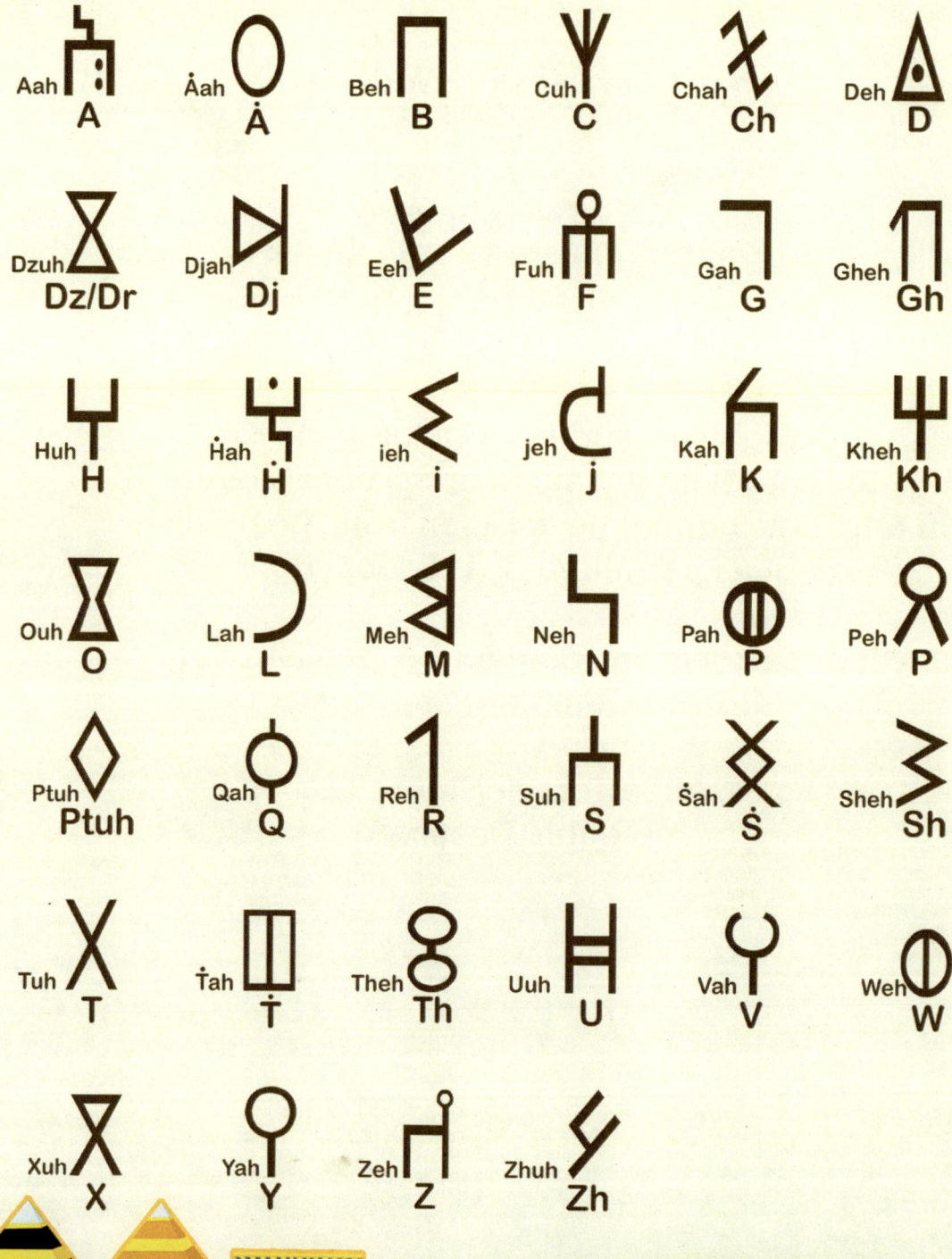

Lah'aj Kids
DIVISION WORKSHEETS WITH ṠABAṪIYYA

ṠABAṪIYYA RAQAM·AAT — SABAEIC NUMBERS

Symbol	Value	Name
⊙	0	Safu
│	1	Wahu
∧	2	Athu
⅄	3	Thalu
+	4	Rabu
✕	5	Khamu
✳	6	Satu
✳	7	Sabu
✳	8	Tamu
⊛	9	Tasu
∩	10	Ashu
∩│	11	Ashu Wu Wahu
∩∧	12	Ashu Wu Athu
∩⅄	13	Ashu Wu Thalu
∩+	14	Ashu Wu Rabu
∩✕	15	Ashu Wu Khamu
∩✳	16	Ashu Wu Satu
∩✳	17	Ashu Wu Sabu
∩✳	18	Ashu Wu Tamu
∩⊛	19	Ashu Wu Tasu
∩∩	20	Athura
∩∩│	21	Athura Wahu
∩∩∧	22	Athura Athu
∩∩⅄	23	Athura Thalu
∩∩+	24	Athura Rabu
∩∩✕	25	Athura Khamu
∩∩✳	26	Athura Satu
∩∩✳	27	Athura Sabu
∩∩✳	28	Athura Tamu
∩∩⊛	29	Athura Tasu
∩∩∩	30	Thalura
∩∩∩∩	40	Rabura
∩∩/∩∩	50	Khamura
∩∩∩/∩∩∩	60	Satura
∩∩∩/∩∩∩∩	70	Sabura
∩∩∩∩/∩∩∩∩	80	Tamura
∩∩∩/∩∩∩/∩∩∩	90	Tasura

Symbol	Name	Value
ҁ	Mayu	100
ҁҁ	Mayu Mayu	200
𓀠	Afu	1,000
𓂢	Ashu Afu	10,000
𓆼	Mayu Afu	100,000
𓁨	Malu	1,000,000

Lah'aj Kids
DIVISION WORKSHEETS WITH ṠABAṪIYYA

ꜩꜫꜩꜫꜩ · RANAN · NAME _____

ꜩꜫꜩꜫꜩ · KHADUT · DATE _____

_____ /ᑎᑎ✶ · 25

ṠAHAF WAHU · PAGE 1

Lah'aj Kids
DIVISION WORKSHEETS WITH ŚABATIYYA

RANAN · NAME _____ /⊓⊓✶ · 25

KHADUT · DATE _____

S'AHAF ATHU · PAGE 2

Lah'aj Kids — DIVISION WORKSHEETS with Šabatiyya

ᛒᛆᛒᛆᚾ · RANAN · NAME

ᚷᛘᛆᚺᛏ · KHADUT · DATE

/ᑎᑎ✕ · 25

Lah'aj Kids
DIVISION WORKSHEETS WITH ŚABATIYYA

/∩∩⋇·25

▰▰▰▰▰▰▰▰▰▰▰▰▰▰▰▰▰▰▰
ᛊᚺᛊᚺ1 · RANAN · NAME

▰▰▰▰▰▰▰▰▰▰▰▰
ᛪᚻᗄᚻᛃ · KHADUT · DATE

Lah'aj Kids
DIVISION WORKSHEETS WITH ŚABATIYYA

⌐⌐⌐⌐⌐⌐⌐⌐⌐⌐⌐⌐⌐ /∩∩⋇·25

ᛋᚻᛋᚻ1·RANAN·NAME

⌐⌐⌐⌐⌐⌐⌐⌐⌐⌐⌐⌐⌐
ⵝᚻᛋᚻᛍ·KHADUT·DATE

∩ ÷ |
∩∩∩ ÷ ∩|
∩∩⋇ ÷ ∩∧
⊙ ÷ ⊙
⋇ ÷ ⊙

∪Υ ÷ ∪
⋇ ÷ ∧
∩∩⋇ ÷ ∩
⋇ ÷ ⋇
∩∩∧ ÷ +

∩∩∩⋇ ÷ ⊛
∩∩| ÷ ∩|
⋇ ÷ ⋇
∧ ÷ ⊙
∩⋇ ÷ ∩∧

+ ÷ +
∩∩∧ ÷ ⋇
∩∩ ÷ ∩∧
⊛ ÷ ⋇
∩| ÷ ⊛

Υ ÷ |
∩∧ ÷ ⋇
∩∩∩∩⊛ ÷ ⋇
⋇ ÷ |
⋇ ÷ ⋇

S'AHAF KHAMU · PAGE 5

Lah'aj Kids — DIVISION WORKSHEETS WITH ŠABATIYYA

/∩∩⋇·25

ᓀᕐᕐᔭ·RANAN·NAME

ⵝᐱᐊᐡᵾ·KHADUT·DATE

∩⋇ ÷ ⋇	∩∧ ÷ ―	⋇⋇ ÷ ―	∩⎓ ÷ ⵝ	⊛ ÷ ✝
∩ ÷ ⋇	⊛ ÷ ⵝ	∩ʏ ÷ ∩ʏ	∩ ÷ ∩	∩⎓ ÷ ⋇
∩ ÷ ∩	∩∩∧ ÷ ∩⎓	⋇ ÷ ―	∧ ÷ ―	∩∧ ÷ ∧
∩∩∩ ÷ ∩⎓	✝ ÷ ⋏	⋇ ÷ ⋏	― ÷ ⊙	⋇ ÷ ∧
∩∩∧ ÷ ⊛	∩ ÷ ∧	⊛ ÷ ✝	⋇ ÷ ⋇	∩∩∩⎓ ÷ ∩⎓

ⴲⵝⵓⴰ ⵚⴰⵝⵓⴲ · S'AHAF SATU · PAGE 6

Lah'aj Kids — DIVISION WORKSHEETS with Šabatiyya

RANAN · NAME

KHADUT · DATE

Lah'aj Kids
DIVISION WORKSHEETS WITH ŠABAṪIYYA

⋔ᛞᛞ⋇ · 25

_____
ᛙᚺᛙᚺᛐ · RANAN · NAME

_____
ᚷᛮᚼᛮᚾ · KHADUT · DATE

Lah'aj Kids
DIVISION WORKSHEETS WITH ṢABATIYYA

/ⵑⵑ⋇ · 25

⸰⸰⸰⸰⸰ · RANAN · NAME

⸰⸰⸰⸰⸰ · KHADUT · DATE

ṢAHAF TASU · PAGE 9

Lah'aj Kids — DIVISION WORKSHEETS WITH ṠABAṪIYYA

RANAN · NAME

KHADUT · DATE

Lah'aj Kids
DIVISION WORKSHEETS WITH ŠABATIYYA

⎯⎯⎯⎯⎯⎯⎯⎯⎯⎯⎯⎯⎯⎯⎯⎯⎯⎯
ᚱᚨᚾᚨᚾ · RANAN · NAME

/ᑎᑎ✶ · 25
⎯⎯⎯⎯⎯⎯⎯⎯⎯⎯⎯⎯⎯⎯⎯⎯⎯⎯

⎯⎯⎯⎯⎯⎯⎯⎯⎯⎯⎯⎯⎯⎯⎯⎯⎯⎯
ᚷᚺᚨᛞᚢᛏ · KHADUT · DATE

$\dfrac{∩∧}{∩}$ $\dfrac{✶}{+}$ $\dfrac{∩∧}{∧}$ $\dfrac{∧}{∧}$ $\dfrac{∩}{⊛}$

$\dfrac{∩∧}{∩}$ $\dfrac{∩}{✶}$ $\dfrac{∩∩⊛}{∩∣}$ $\dfrac{✶}{∧}$ $\dfrac{✶}{+}$

$\dfrac{∩∩+}{✶}$ $\dfrac{∩✶}{✶}$ $\dfrac{∩∩✶}{✶}$ $\dfrac{∩∧}{∩∣}$ $\dfrac{∩∩✶}{∩∣}$

$\dfrac{⊛}{∧}$ $\dfrac{∩∣}{∩}$ $\dfrac{⊛}{⋏}$ $\dfrac{∩∣}{∩∣}$ $\dfrac{∩✶}{∩}$

$\dfrac{∩+}{⊛}$ $\dfrac{∩✶}{⊛}$ $\dfrac{∩∧}{✶}$ $\dfrac{∩∣}{∩}$ $\dfrac{✶}{+}$

SAHAF ASHU WAHU · PAGE 11

Lah'aj Kids
DIVISION WORKSHEETS WITH ŠABATIYYA

⊹ᛘ⊹ᛘꔷ • RANAN • NAME

⅄ꔷ⌂ᛘꔷ • KHADUT • DATE

/∩∩⋇ • 25

Lah'aj Kids
DIVISION WORKSHEETS WITH ŠABATIYYA

RANAN • NAME

KHADUT • DATE

Lah'aj Kids
DIVISION WORKSHEETS WITH ṠABAṪIYYA

/∩∩⋇·25

ᚳᚪᚳᚪᚱ · RANAN · NAME

ᚷᛘᚪᚾᛋ · KHADUT · DATE

Lah'aj Kids
DIVISION WORKSHEETS WITH ṠABATIYYA

/ⵐⵐ✕ · 25

ᚱᚨᚾᚨᚾ · RANAN · NAME

ᚲᚺᚨᛞᚢᛏ · KHADUT · DATE

ṠAHAF AŠHU KHAMU · PAGE 15

Lah'aj Kids — DIVISION WORKSHEETS WITH ŚABAṪIYYA

‾‾‾‾‾‾‾‾‾‾‾‾‾‾ /ℼℼ✕·25
ꝗꝑꝗꝑ1·RANAN·NAME

‾‾‾‾‾‾‾‾‾‾‾‾‾‾
ẊⱧꝕⱧꝓꝪ·KHADUT·DATE

∩∩ ÷ ⋏	✹ ÷ ✹	∩∩✳ ÷ ∩⋏	✳ ÷ ✳	∩∩+ ÷ ✹
✳ ÷ ✳	∩⋏ ÷ ∩⋏	✳ ÷ ✳	⋏∩ ÷ ✹	+ ÷ ⋏
∩∩+ ÷ ✳	∩⋏ ÷ ✳	∩⋏ ÷ ∩⋏	✳ ÷ ✳	✳ ÷ ⋏
∩∩∩✳ ÷ ✹	∩⋏✳ ÷ ∩⋏	✳ ÷ ✳	∩⋏ ÷ ∩⋏	∩∩+ ÷ ∩⋏
∩∩+ ÷ ∩⋏	∩⋏ ÷ ✳	⋏ ÷ ⋏	✳ ÷ ✳	✳ ÷ 1

SAHAF ASHU SATU · PAGE 16

Lah'aj Kids
DIVISION WORKSHEETS WITH ŚABATIYYA

⸽⸽⸽⸽⸽⸽⸽⸽⸽⸽ · RANAN · NAME

⸽⸽⸽⸽⸽⸽⸽⸽⸽⸽ · KHADUT · DATE

⸽⸽⸽⸽⸽⸽⸽⸽⸽⸽ /∩∩⁕ · 25

 · SAHAF ASHU SABU · PAGE 17

Lah'aj Kids
DIVISION WORKSHEETS WITH SABATIYYA

_____
ꓳꓶꓶꓵ · RANAN · NAME

_____ /ꓵꓵx · 25

_____
XHAꓛꓲꓴ · KHADUT · DATE

∩∧ ÷ ∩|		÷		∩∧ ÷ ⁕	⊛ ÷ ⁕	∩∩| ÷ ∩|
⊛ ÷ ⁕	∩⊛ ÷ ∩∧	∩∧ ÷ ∩∧	U ÷ Y	∩+ ÷ ⊛
x ÷ ∧	x ÷ Y	∩| ÷ ⊙	+ ÷ +	∩x ÷ x
∩x ÷ ⁕	⁕ ÷ x	⁕ ÷ Y	⊛ ÷		∩	÷ ⊙
⊛ ÷ +	∩ ÷ ⁕	∩| ÷ +	∩| ÷ Y	⁕x ÷ x

Lah'aj Kids
DIVISION WORKSHEETS WITH ŠABATIYYA

ꝩꝩꝩ · RANAN · NAME

ꝩꝩꝩ · KHADUT · DATE

/ꝩꝩ✕ · 25

SAHAF ASHU TASU · PAGE 19

Lah'aj Kids — DIVISION WORKSHEETS with Šabaṫiyya

⸗⸗⸗⸗⸗⸗⸗⸗⸗⸗⸗⸗⸗⸗⸗⸗ /⋂⋂✕ · 25
ᚾᚷᚾᚷ1 · RANAN · NAME   ⸗⸗⸗⸗⸗⸗⸗⸗⸗⸗⸗⸗⸗⸗⸗⸗⸗⸗

⸗⸗⸗⸗⸗⸗⸗⸗⸗⸗⸗⸗⸗⸗⸗⸗⸗⸗⸗⸗
ᚷᚻᚾᚷᛈᛈ · KHADUT · DATE

Lah'aj Kids
DIVISION WORKSHEETS WITH ṠRBAṪIYYA

/∩∩✵ · 25

ᚴᚬᚴᚬ1 · RANAN · NAME

ፘᚻᚪᚼᛡ · KHADUT · DATE

SAHAF ATHURA WAHU · PAGE 21

Lah'aj Kids
DIVISION WORKSHEETS WITH ṠABATIYYA

_____ /ⴖⴖ⋇ · 25
ꓓꓯꓳꓯꓓ · RANAN · NAME

_____
ⵝꓱꙭꓴ · KHADUT · DATE

Lah'aj Kids
DIVISION WORKSHEETS with Šabaṭiyya

さささ1 • RANAN • NAME

メ日ムキヤ • KHADUT • DATE

Lah'aj Kids
DIVISION WORKSHEETS WITH ŠABATIYYA

/∩∩✷·25

ᛣᚼᛋᚼᛀ · RANAN · NAME

ᚼᛘᚼᛘᛀ · KHADUT · DATE

ᛘᚼᛋ1ᛋ1ᛘᛀᛋ ᛘᛋᛀᛣᛘ · SAHAF ATHURA RABU · PAGE 24

Lah'aj Kids
DIVISION WORKSHEETS WITH ṠABATIYYA

NOTES

Lah'aj Kids
DIVISION WORKSHEETS WITH ŚABATIYYA

NOTES

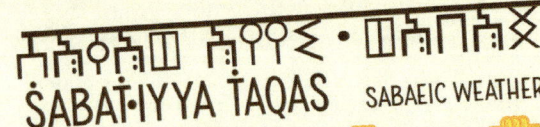

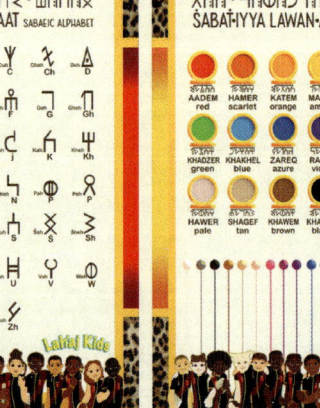
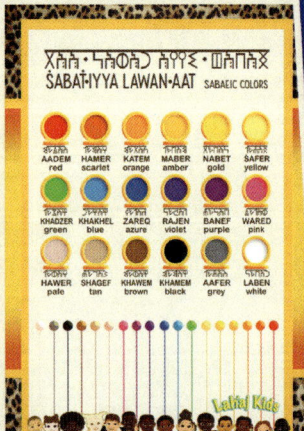

FIND MORE AT TEMPLEBABIES.COM

NEW ŚABAṮIYYA POSTERS
large 12" x 18"

Master S'abat'iyya like never before, with our new language posters! Teaching the fundamental vocabulary for time, numbers, colors, conjugations & more! Perfect to hang in your homeschool space, office, living room, or wherever suits your education. For all ages eager to learn.

TEMPLEBABIES.COM — a place for people like me!

Made in the USA
Middletown, DE
15 November 2025

21453271R00018